YOUR KNOWLEDGE HAS VALUE

Bibliographic information published by the German National Library:

The German National Library lists this publication in the National Bibliography; detailed bibliographic data are available on the Internet at http://dnb.dnb.de .

Imprint:

Copyright © 2019 GRIN Verlag
Print and binding: Books on Demand GmbH, Norderstedt Germany
ISBN: 9783346003041

This book at GRIN:

https://www.grin.com/document/492993

Morteza Valiollahpur

Scientific Breakthrough. Unprecedented Research Methods

GRIN Verlag

GRIN - Your knowledge has value

Since its foundation in 1998, GRIN has specialized in publishing academic texts by students, college teachers and other academics as e-book and printed book. The website www.grin.com is an ideal platform for presenting term papers, final papers, scientific essays, dissertations and specialist books.

Visit us on the internet:

http://www.grin.com/

http://www.facebook.com/grincom

http://www.twitter.com/grin_com

Title

Scientific Breakthrough

" Unprecedented Research Methods "

Author

Morteza Valiollahpur, Ph.D.

Academic Elite

Table of Contents

Gift to

God, Creator of the Universe

PREFACE

In order to explain plans, strategies, experimental research methods, strategies, uncertainties ... (from Theories, Philosophies, Explanations, Effects, Aftereffects, Comparisons, Frameworks, Models, Share, Reasons, The share of aid, ... to Phenomenology, Ontology, Epistemology, and ... as a small part of research) the important new thinking in the research method of sciences. The scientific breakthrough is a situation formed on the perception of perimeter of a researcher who has a insight. This status show self face in trade, invest, culture, politics, economics, society and etc. A method of science that it has been given a appropriate answer in the past 50 decades in international scientific societies, but has not been given a response to its evolution. It's mean in that we should thinking and understand the true science of the research. For all of the issues that you are designing, the revealed structures will justify most organized and unstructured methods that it will show you the right policies for using information. How when that you wish assess the target a micro level for example for an organization or at a macro level for example for the community to give the best result to yourself, but when you use your essential information, only the result is enough, in other words, do not include them in your attitude systems the data that references the necessary information to you does not bring you trusts. it may also show deviations from your values so that you think moving with you. This subject goes beyond the scientific process of countries and when it becomes clear that you want with all sorts of your reason to engaged in analyze the complexities of science in ways beyond of linear and nonlinear thinking. The scientific breakthrough is a situation formed on the perception of perimeter of a researcher who has a insight. This status show self face in trade, invest, culture, politics, economics, society and etc. A method of science that to it has been given a appropriate answer in the past 50 decades in international scientific societies, but has not been given a response to its evolution. It's mean in that we should thinking and understand the true science of the research. Looking at the developmental indicators the entry into the new era placed structures in self that have embodied non-realistic practices with the principles of realistic, yet at with simple outward but complex inward that can not be easily overlooked and these are not just the principles of mathematics or the basic fundamentals of knowledge. cinema, computer games, Information Technology to solve overnight operations to books, research projects and papers on the bases of scientific as Science Direct, Elsevier, Wiley, Emerald, Springer, PupMed, Institute of Electrical and Electronics Engineers, Researchgate, Sage, Eric and ... due to the circumstances of the new era has emerged new fields.

Morteza Valiollahpur

April, 2018

Important: This statement for the first time was presented by Dr. Morteza Valiollahpur in April 2018 at the Iranian National Foundation of Elites at: http://en.bmn.ir.

Chapter One

Breakthrough Imagination

1-1. Simple Appearance of Complex Analyzes ◄---► Effective Processes

Always A small number of factors have the approximate ability that all the information obtained by a larger set of variables, be simplify in easy structures derived from the result of describing complex features. In other words the number of asked factors (whatever), produce variables with the implication of a logical construct, so that the description of the explicit and hidden agents is not only easier but more logical and clearer. Integration of the balanced dataset sets, justify the minimum factors for the hypothetical or theoretical structures. Therefore useful test of interpretation massive of maximum savings in factors explaining logical constructions. Experiments each time a factors applies special conditions, so that when there are interactions for each factor, in order to avoid misleading results, it is necessary to estimate the effects of the other at several other levels, which obtained results in the range of experimental conditions. The pattern of action for conducting an experiment (any action that leads to the collection of information) so that achieved by using observational and statistical methods to validate the conclusion, it make possible to achieve the factors. Be careful that the controllable factors are controlled by selecting any test (whatever). and the uncontrollable factors also trying to minimize changes by selecting the right model and hypothesis on the experimental observations. If we seek to investigate the relationship between a nominal or sequential variable or a distance or relative in a given experiment, can we claim that changes in the nominal or sequential variable, or distance or relative variables, can justify the modeling experiment. Always possible to reduce the structure of a large number of variables into a smaller number of hidden dimensions. Its main purpose is to adhere to the principle of saving through the use of the smallest explanatory concepts in order to explain the maximum results of the experiment. In other words, The least the underlying factors of variables can be used to explain complex phenomena and other variables in the possible experiment framework are related to the interactions of designed condition, and the probability infinite, future variables is the result of their sharing in these seemingly structured factors of the variables pattern. Always the invisible factors are justifiable by set of visible variables. It is necessary to explain that in a huge experiment all the justifiable factors for the amount of effect and comparison with each other are always discussed. The discussion of new methods enter new complexities to innovation structures. There are several content and concepts of conventional correlations (communication between variables) in mathematics and engineering sciences, natural sciences, humanities and social sciences. A natural phenomenon with any approximate can be a standard condition for developmental structures. Often, the interest in examining the relationship between a group of variables in the analysis of linear combinations of the two groups with the aim of using the nonlinear states that involved random in it, based on theories of random trials are justifiable, however, a few of factors should have the ability to provide all the information provided by a larger set of variables. Providing a mathematical description of our beliefs about the systematic properties of a random phenomenon is the first step in this kind of analyzing. It is necessary make key idea in variables whose measurement is a detailed description of the logical outcome of a random phenomenon. Providing the background for implementing other scientific methods in advancing other sciences is the beginning of an analysis of the real context for research to discover something about testable processes. Always creating a optimum condition with having increase status rational instead of using a cause-and-effect logical process based on experimental operation leads to the discovery of new options related to justifiable structures.

1-2. Factor Experiment

In a one-factor test, which is a situation of one-mode situations, a factor has "a" different level that aims to compare them with each other. And the observed response from each of a level is a random variable. According to formula (1-1), describe observations of the statistical linear model;

$$y_{ij} = \mu + \tau_i + \varepsilon_{ij} \qquad\qquad i=1,2,\ldots, \ a \ , \quad j=1,2,\ldots,n$$
(1-1)

(Source: C. Montgomery translation by Shahkar, PP. 61)

But in a test of 2^2 which is a two-mode condition, two factors have "a" and "b" different levels that aimed comparing them with each other. The geometric display of the high and low levels of the factors a and b that evident in Fig. (1-1). And again, the observed response from each of the a and b levels is a random variable.

Fig 1-1. Structural geometric disp

(Source: C. Montgomery Translation of Shahkar, PP. 318)

In these experiments, the subject is not summarized in two-mode because we are faced with 2^k factors. And for more than 2 observations, we encounter with similar structures in three-mode. The description of the observations of the statistical linear model of these designs is based on the more parameters of the relation (1-1) as follows.

$$y_{ij} = \mu + \tau_i + \beta_j + (\tau\beta)_{ij} + \varepsilon_{ij} \qquad\qquad i=1,2,\ldots, \ a \ , \quad j=1,2,\ldots,n$$
(1-2)

Also, the subject is different for testing more than three and several levels. When faced with an experiment 3^3, it means that each factor is in three levels and 27 combinations with 26 degrees of freedom, each of the main effects have two degrees of freedom, and any two-factor interaction, 4 degrees of freedom, and any three-factor interaction, 8 degrees of freedom. The geometric shape of such a test is shown in Fig. (1-2). And again, the observed response from each of a and b and c is the level of a random variable.

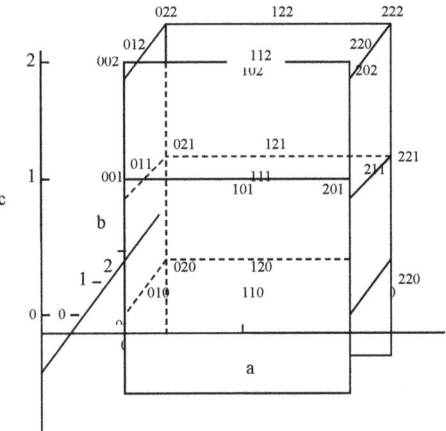

Chart 1-2. Structural geometric display at three levels

(Source: C. Montgomery Translation of Shahkar, PP.459)

In this case, we encounter with 3^k functional structures. The description of the observations of the statistical linear model of these designs is based on the more parameters of the relation (1-1) as follows.

$$y_{ij} = \mu + \tau_i + \beta_j + \gamma_k + (\tau\beta)_{ij} + (\tau\gamma)_{ik} + (\beta\gamma)_{ik} + (\tau\beta\gamma)_{ijk} + \varepsilon_{ijk} \qquad i=1,2,\ldots,a \; , \; j=1,2,\ldots,n \qquad (1\text{-}3)$$

1-3. A Simple Sample

We assume that in the specialized field of humanities in the branch of business management, in a state of strategic situation, we ask a question. A fast food of marketing test has put three new menu items on the east coast and west coast of the United States. To find the same popularity, 12 Franchisee restaurants have been randomly selected from each coast and selected to participate in the study. In accordance with the great design of the market test in 12 coastal restaurants in the east, 4 restaurants for the first menu item selected randomly, 4 other restaurants for the second menu items and 4 remaining restaurants selected for the last item in the menu. The 12 western beach restaurants are also arranged same way. Management is determined to decide on the difference between the average sales volume from the two coastal areas.

Probably a specialist in industrial management from the human sciences branch suggests the use of decision tree methods, AHP and ANP or fuzzy patterns. An industrial engineering specialist from the Engineering Sciences Branch provides similar structures for using the same methods of advanced operations research. An expert on computer games from the field of art will offers the use of strategic complexity. And a software engineer or computer programmer engineer will be able to demonstrate the programmer complexity. A trained psychologist or farmer engineer suggests using experimental designs. But you can claim that the real and unreal mystery a clear-cut problem with a clear solution for every manager and specialist and policymaker at the macro level of any society, has placed logical and irrational conditions in their respective patterns. The human imagination always Includes widest and reproduces scenes. From the viewpoint of thinkers and theorists, the understanding of symbols does not mean the discovery of previously existing knowledge other than the patterns and structures of our universe in the sense of capturing processes that at any cross-section time different take many forms and usually have ambiguity. Let's see an example:

Suppose the following table represents the face of the 3 new menu items after one week of marketing testing. The left half of the table represents sales figures from three east coastal restaurants. The right half represents the west coast restaurant. At level 0.05 check whether the average volume of new item menu items is equal. Also decide if the average sales volume varies from two coastal areas.

East Coast:
==========

Item1	Item2	Item3	
E1	25	39	36
E2	36	42	24
E3	31	39	28
E4	26	35	29

West Coast:
==========

Item1	Item2	Item3	
W1	51	43	42
W2	47	39	36
W3	47	53	32
W4	52	46	33

Save the data in txt format:

	Item1	Item2	Item3
E1	25	39	36
E2	36	42	24
E3	31	39	28
E4	26	35	29
W1	51	43	42
W2	47	39	36
W3	47	53	32
W4	52	46	33

Load the data with the read.table command in the R-Programming software:

< df3 = read.table(file = "G:/fastfood-3.txt", header = TRUE)

< df3

The output is as follows:

	Item1	Item2	Item3
E1	25	39	36
E2	36	42	24
E3	31	39	28
E4	26	35	29
W1	51	43	42
W2	47	39	36
W3	47	53	32
W4	52	46	33

Define the vector as follows:

```
< r = c(t(as.matrix(df3)))
< r
```

The output is as follows:

[1] 25 39 36 36 42 24 31 39 28 26 35 29 51 43 42 47 39 36 47 53 32 52 46 33

Determining new variables for the first level and the number of observations:

```
< f1 = c("Item1", "Item2", "Item3")
< f2 = c("East", "West")
< k1 = length(f1)
< k2 = length(f2)
< n = 4
< tm1 = gl(k1, 1, n*k1*k2, factor(f1))
```

< tm1

The output is as follows:

[1] Item1 Item2 Item3 Item1 Item2 Item3 Item1 Item2 Item3 Item1 Item2 Item3 Item1 Item2 Item3 Item1 Item2 Item3 Item1 Item2 Item3 Item1 Item2 Item3

Levels: Item1 Item2 Item3

Similarly, a new vector for the second level:

< tm2 = gl(k2, n*k1, n*k1*k2, factor(f2))

< tm2

The output is as follows:

[1] East East East East East East East East East East East East West West West West West West West West West West West West

Levels: East West

The aov command that describes r-response by factors for two levels tm1 and tm2:

<av = aov(r ~ tm1 * tm2)

summary order for printing ANOVA table:

< summary(av)

The output of the statistical software the R-Programming is as follows:

	Df		Sum	Sq Mean	Sq F	value Pr(>F)
tm1	2		385	193	9.55	0.0015 **
tm2	1		715	715	35.48	1.2e-05 ***
tm1:tm2	2		234	117	5.81	0.0113 *
Residuals	18	363	20			

Since the value of P-0.0015 for menu items is less than the 0.05 level, the zero-hypothesis is rejection and the average of the volume of sales of new menu items is equal. In addition, P-1.2e-05 is less than the 0.05 level than the east and west coasts. This indicates that there is a significant difference in the total volume of sales between the beaches. Finally, P-0.0113 (<0.05) shows that there is interaction between menu items and coastal locations. for example, for customers in different coastal regions that have different tastes.

1-4. Recognition, Discussion

The discussion of the breakthrough scientific method is complex. Human-science research is a quasi-practical theory. There, we do not have a theory about each issue, but we are faced with a wide range of theories, all of which are somewhat flawed in contrast to other theories. No matter how comprehensive the theory is, there's something we did not see because we are dealing with the human being and the Its value system. Evidence has shown that the world is breaking the paradigms of modernity. Situations are no longer systematic and easy, and issues that have effective on different theories in each its aspect have a complex discipline for new academics that a different theoretical structure of discipline-centered. This is related to the action and the choice of the lateral, contrary to the theory that confronts the net-knowledge. Its methods, in contrast to the traditional methods, do not result in accurate results at present, but provide better conditions for the rational understanding of the current world. The lateral thought is thought to be creative thinking against traditional thought. Person in traditional thinking, one usually makes use of logic within the framework of the assumptions of environmental conditions and constraints, and does not have the ability to analyze other angles of situations, circumstances, situations, and surroundings. But lateral thinking, by eliminating constraints, teaches to people that break the tradition in the thought of solving problems and analyzing the situation. And they seeing other boundary of thought and the another angles of problems. Therefore, we must upgrade the ability to choose and act by non-traditional thoughts. And when this situation is put into effect it will be achieve improve. Reassuring mathematics and statistics for vast applications in interacting and interconnecting different parts of science with each other that interact and react, and at present, no theories and meta-theories can cover this complexity, and that it should include all the dimensions and characteristics, until be defendable and its predetermined features show their willingness in the tools used and put the structures in the process cycle and subject, content, theory, data, information, analysis, and ... accepted under pivotal process of effectiveness And prior to the final selection, based on the process of the decision of mind arranges the explanation of the theory organizes the theory's explanation, from the scheduled route-extraction process Logical generalizations arise from the underlying phenomena and given the fundamental steps of a scientific inquiry, it is clear that the vast domain of all scientific research with the logical structuring of mind tied to many sciences. The relationships within the systems are summarized in a variety of management models which are an important tool for explaining the phenomena observed in the systems. Therefore, modeling is a tool for understanding the complex relationships in the evolution of the conditions governing the research variables and can be employed in the management of effective change management. The systematic perspectives include the model of complex systems in the innovation management systems and formulate the conceptual form of the model and its structural algorithm and quantify, evaluate, and apply it in a systematic analysis from a managerial point of view. Thus, at the stage of formulating the conceptual form of the model, based on the objectives of the systematic analysis of the model, we will deal with ecosystems which are the actual components of the decision-making model. The interactions of different complex variables lead to the development of different branches of science in accordance with the demands of society, circumstances, and time developments. Considering that the innovative performance of each country is determined by its national innovation system and due to the universe developments, the conditions should be arranged in such a manner that the priorities of the management system are set by guaranteeing sustainable development and explained on a wider scale by complex structures. It is necessary to structure the priorities of the management system for innovation and form the substrates consistent with it. By considering the time requirements, capacities, and public awareness of the process of growth, development, and progress of the country, the components of innovation management models will improve by promoting the competitiveness in any country. Furthermore, enhancing the quality and standards of living, ensuring profitable relationships, human resource development, and social development will be reflected in the process cycle of sustainable development standards. All phenomena are part of a social experiment with using the marvel of the wonderful phenomena of our era. an idea in mind, a simple thought with all the complexities and perceptions of world realities and non-realities can create a source of change. The existence of rules based on simple rules with reversible complexities for pass from critical situation is based on breaking the truth. A reasonable way to better balance for understand the current universe is the designing affairs without rules in order to control the situation. But aligning with the illogical path is a principles for realizing the rational truth of life in society. disrupt the ruling conditions by putting together order and chaos in order to achieve the desired goal rather than returning to the previous time in order to get rid of the previous learning of the result of our experiences, creates better condition. At this time, everything will show tendency to balanced condition. Moving from one state to another, optimum in our vision, in addition to extraordinary Ingenuity, requires the discovery of an unreal real nature. Non-realities that appear in the simplest information. Understanding the inward of essential information before doing any action, justify most of the organized and unorganized methods that it will reveal the path to the correct use of information policies. For the problems of the current world the structures need to examine the most minor aspects of experimental information, and considering this subject in attitudinal systems lead to bring trustworthy the referrals data of necessary information in accordance with values that moved in the right direction with the our demands. I draw your attention to the above example with R-Programming.

Chapter Two

Breakthrough

Research Methods

2-1. New Way of Thinking in Research Methods

In order to explain plans, strategies, experimental research methods, strategies, uncertainties ... (from Theories, Philosophies (Idealism, Realism, Pragmatism, Existentialism, etc.), Explanations, Effects, Aftereffects, Comparisons, Frameworks, Models, Share, Reasons, The share of aid, ... to Phenomenology, Ontology, Epistemology, and ... as a small part of our research) the important new thinking in the research method of sciences. This status show self face in politics, politics, economics, society, culture, and etc. A methodology that to it has been given a appropriate answer in the past 50 decades in international scientific societies, but has not been given a response to its evolution. Sometimes a motion in the mind can lead to the expression of an evolutionary methodology or standard methodology. I draw your attention to the following method (Presented from the Douglas C.Montgomery Books by Morteza Valiollahpur, Elite Paper-2014).

2-2. Thinking in Research Methods

Douglas C. Montgomery, Suppose we have "a" treatments or different levels of a single factor that we wish to compare. The observed response from each of "a" treatments is a random variable. The data would appear as in table 1. An entry in table 1 (e.g., yij) represents the jth observation taken under factor level or treatment i.

Table 2-1. Typical Data for a Single-Factor Experiment

Treatment (Level)	Observations						Totals	Averages
1	y_{11}	y_{12}	...	y_{1n}	$y_1.$	$\bar{y}_1.$		
2		y_{21}	y_{22}	...	y_{2n}	$y_2.$	$\bar{y}_2.$	
.	.	.	.	...	.	.	.	
.	.	.	.	...	.	.	.	
a	y_{a1}	y_{a2}	...	y_{an}	$y_a.$		$\bar{y}_a.$	
	$y_{..}$	$\bar{y}_{..}$						

Source: Design and Analysis of Experiments, Eighth Edition, Douglas C. Montgomery, PP. 68

Models for the data, we will find useful to describe the observations from an experiment with a model. One way to write this model is:

$$y_{ij} = \mu + \tau_i + \varepsilon_{ij} \qquad\qquad i=1,2,\dots, a \ , \quad j=1,2,\dots,n$$
(2-1)

Table 2-2. Parameter

μ (Total mean), τ_i (Effect of i-th treatment), ε_{ij} (Random error)

Parameter Definition

According to Table 2-1 and Relationship (1), yij (ij) observation, μ common parameter for all treatments called total mean, τ_i the common parameter is i-th treatment called the effect of i-th treatment, and ε_{ij} is a random error component. We will see descriptions of observations of the statistical linear model;

$$y_{ij} = \mu + \tau_i + \varepsilon_{ij} \qquad\qquad i=1,2,\ldots, a \ , \quad j=1,2,\ldots,n$$
(1)

Equation 2-1 can written as follows; that is, $E(y_{ij}) \equiv \mu_i = \mu + \tau_i, i = 1, 2, \ldots, a;$

$$y_{ij}=\mu_i \qquad\qquad\qquad\qquad\qquad\qquad\qquad + \varepsilon_{ij}$$
(2-2)

In this form of the model, where y_{ij} is the ijth observation, μ_i is mean, and ε_{ij} is a random error. We are interested in testing the equality of the a treatment means; the appropriate hypotheses are;

$H_0: \mu_1 = \mu_2 = \ldots = \mu_a$

$H_A: \mu_i \neq \mu_j \quad$ (i,j) برای حداقل یک جفت

Table 2-3. The Analysis of Variance Table for the Single-Factor

Source of Variation	Sum of Squares	Degrees of Freedom	Mean Squares	F_0
Between treatments	SSTr	a - 1	MSTr = SSTr/(a-1)	F_0 = MSTr/MSE
Error (within treatments)	SSE	N - a	MSE = SSE/(N-a)	
Total	SST	N - 1	-	-

Source: Design and Analysis of Experiments, Eighth Edition, Douglas C. Montgomery, PP. 75

In table 2-3 F_0; Fisher Statistic, SSTr; Sum of Squares for Treatments, SSE; Sum of Squares for Error, SST; Total Sum of Squares, MSTr; Mean Square for Treatments, MSE; Mean Square for Error, N is total number of observations. The name analysis of variance is derived from a partitioning of total variability into its component parts. It states that the total variability in the data, as measured by the total corrected sum of squares, can be partitioned into a sum of squares of the differences between the treatment averages and the grand average plus a sum of squares of the differences of observations within treatments from the treatment average. Therefore;

$$SST = SSTr + SSE \quad , \quad N-1 = a-1 + N-a \quad , \quad N = an \qquad\qquad (2-3)$$

According to Table 1, the equality Equation (2-3) is as follows;

$$\sum_{i=1}^{a}\sum_{j=1}^{n}(y_{ij}-\bar{y}_{\bullet\bullet})^2 = n\sum_{i=1}^{a}(\bar{y}_{i\bullet}-\bar{y}_{\bullet\bullet})^2 + \sum_{i=1}^{a}\sum_{j=1}^{n}(y_{ij}-\bar{y}_{i\bullet})^2 \tag{2-4}$$

We see that the "dot" subscript notation implies summation over the subscript that it replaces. In Equation (2-4) $\bar{y}_{i\bullet} = y_{i\bullet}/n$, $\bar{y}_{\bullet\bullet} = y_{\bullet\bullet}/N$. the degrees of freedom for SST and SSE add to N-1, the total number of degrees of freedom. Therefore, if the null hypothesis of no difference in treatment means is true, the ratio;

$$F_0 = \frac{SS(Tr)/(a-1)}{SSE/(N-a)} \triangleq \frac{MS(Tr)}{MSE} \tag{2-5}$$

We should reject H_0 on values of the test statistic that are too large. This implies an upper-tail, one-tail critical region. Therefore, we should reject H0 and conclude that there are differences in the treatment means if;

$$F_0 > F_{\alpha,\, a-1,\, N-a} \tag{2-6}$$

Where F_0 is computed from equation 2-5 alternatively, we could use the P-value approach for decision making.

Now, we explain the basic procedure analysis of covariance for a single factor experiment with one covariate. Assuming that there is a linear relationship between the response and the covariate, we find that an appropriate statistical model is;

$$y_{ij} = \mu + \tau_i + \beta(x_{ij} - \bar{x}_{\bullet}) + \varepsilon_{ij} \qquad\qquad i=1,2,\ldots, a\ ,\quad j=1,2,\ldots,n \tag{2-7}$$

where y_{ij} is the jth observation on the response variable taken under the ith treatment or level of the single factor, x_{ij} is the measurement made on the covariate or concomitant variable corresponding to y_{ij} (i.e., the ijth run), $\bar{x}_{\bullet}$ is the mean of the x_{ij} values, μ is an overall mean, τ_i is the effect of the ith treatment, β is a linear regression coefficient indicating the dependency of y_{ij} on x_{ij}, and ε_{ij} is a random error component. Notice from equation 2-7 that the analysis of covariance model is a combination of the linear models employed in analysis of variance and regression. Since parameter μ is preserved as the overall mean, appropriate statistical linear model;

$$y_{ij} \quad = \quad \mu' \quad + \tau_i \quad + \beta x_{ij} \quad + \varepsilon_{ij} \tag{2-8}$$

That in it $\mu' = \mu - \beta\bar{x}_{\bullet}$. Manual computations are usually displayed in an analysis of covariance table such as Table 4 This layout is employed because it conveniently summarizes all the required sums of squares and cross products as well as the sums of squares for testing hypotheses about treatment effects.

Table 2-4. Analysis of covariance for a single-factor experiment with one covariate

Source of variation	Degrees of freedom	Sums of squares and products x	xy	y	Adjusted for regression y	Degrees of freedom	Mean square
Treatments	$a-1$	T_{xx}	T_{xy}	T_{yy}	-	-	-
Error	$a(n-1)$	E_{xx}	E_{xy}	E_{yy}	$SS_E = E_{yy} - (E_{xy})^2/E_{xx}$	$a(n-1)-1$	$MS_E = SS_E/a(n-1)-1$
Total	$an-1$	S_{xx}	S_{xy}	S_{yy}	$SS'_E = S_{yy} - (S_{xy})^2/S_{xx}$	$an-2$	-
Adjusted	-	-	-	-	SS'_E-SS_E	$a-1$	$SS'_E-SS_E/a-1$

Source: Design and Analysis of Experiments, Eighth Edition, Douglas C. Montgomery, PP. 659

Note that, in general, $S + T = E$, where the symbols S, T, and E are used to denote sums of squares and cross products for total, treatments, and error, respectively. The sums of squares for x (concomitant variable) and y (response variable) must be nonnegative; however, the sums of cross products (xy) may be negative.

$$S = T + E \qquad , \qquad an-1 = a(n-1) + a-1 \tag{2-9}$$

Also, in general, the analysis of covariance involves adjusting the observed response variable for the effect of the concomitant. According to sums of squares and cross products table 4 Equality of equation (9) as follows;

$$\sum_{i=1}^{a}\sum_{j=1}^{n}(x_{ij}-\bar{x}_{\bullet\bullet})^2 = \sum_{i=1}^{a}(\bar{x}_{i\bullet}-\bar{x}_{\bullet\bullet})^2 + \sum_{i=1}^{a}\sum_{j=1}^{n}(x_{ij}-\bar{x}_{i\bullet})^2 \quad\rightarrow\quad S_{xx} = T_{xx} + E_{xx} \tag{2-10}$$

$$\sum_{i=1}^{a}\sum_{j=1}^{n}(y_{ij}-\bar{y}_{\bullet\bullet})^2 = \sum_{i=1}^{a}(\bar{y}_{i\bullet}-\bar{y}_{\bullet\bullet})^2 + \sum_{i=1}^{a}\sum_{j=1}^{n}(y_{ij}-\bar{y}_{i\bullet})^2 \quad\rightarrow\quad S_{yy} = T_{yy} + E_{yy} \tag{2-11}$$

$$\sum_{i=1}^{a}\sum_{j=1}^{n}(x_{ij}-\bar{x}_{\bullet\bullet})(y_{ij}-\bar{y}_{\bullet\bullet}) = \sum_{i=1}^{a}(\bar{x}_{i\bullet}-\bar{x}_{\bullet\bullet})(\bar{y}_{i\bullet}-\bar{y}_{\bullet\bullet}) + \sum_{i=1}^{a}\sum_{j=1}^{n}(x_{ij}-\bar{x}_{i\bullet})(y_{ij}-\bar{y}_{i\bullet}) \quad\rightarrow\quad S_{xy} = T_{xy} + E_{xy} \tag{2-12}$$

Assuming no treatment effect for the equation (2-7) the least squares estimators of μ and β shown to form $\hat{\mu} = \bar{y}_{\bullet\bullet}$ و $\beta = S_{xy}/S_{xx}$ that The sum of squares for error in this reduced model is;

$$SS'_E = S_{yy} - (S_{xy})^2/S_{xx} \tag{2-13}$$

20

with an-2 degrees of freedom. According to table 2-4 the difference between $SS'_E - SS_E$ sum of squares with a-1 degrees of freedom. Consequently, to test no treatment effect compute;

$$F_0 = \frac{(SS'_E - SS_E)/(a-1)}{SSE/[a(n-1)-1]} \overset{\Delta}{=} \frac{MSE}{SSE/[a(n-1)-1]}$$

(2-14)

Which, if the null hypothesis is true, is distributed as $F_{\alpha,\ a-1,\ a(n-1)-1}$. Thus, we reject H_0 if;

$$F_0 \qquad\qquad > \qquad\qquad F_{\alpha,} \qquad\qquad a-1, \qquad\qquad a(n-1)-1$$

(2-15)

Where F_0, calculate from equality equation (2-14). Furthermore, for full model equation (2-7), $\beta = E_{xy}/E_{xx}$. Therefore, for assume there is no linear regression relationship (independence xij to yij) compute;

$$F_0 = \frac{(Exy)^2/Exx}{MSE}$$

(2-16)

Thus, we reject H_0 if;

$$F_0 \qquad\qquad > \qquad\qquad F_{\alpha,} \qquad\qquad 1, \qquad\qquad a(n-1)-1$$

(2-17)

Where F_0, calculate from equality equation (2-14).

Chapter Three

Breakthrough

Scientific

3-1. Scientific Breakthrough

The scientific breakthrough is a situation formed on the perception of perimeter of a researcher who has a insight. This status show self face in trade, invest, culture, politics, economics, society and etc. A method of science that to it has been given a appropriate answer in the past 50 decades in international scientific societies, but has not been given a response to its evolution. It's mean in that we should thinking and understand the true science of the research. I draw your attention to a sample than my work.

Table 3-1. Conceptual Framework

Components of A	Components of A	Components of A	Components of A	Components of A	Components of A	Components of A	Components of A	Components of A	Components of A	Components of A	Components of A	Components of A	Conceptual Framework(A) / Conceptual Framework (B)
C113	C112	C111	C110	C19	C18	C17	C16	C15	C14	C13	C12	C11	Components of B
C213	C212	C211	C210	C29	C28	C27	C26	C25	C24	C23	C22	C21	Components of B
C313	C312	C311	C310	C39	C38	C37	C36	C35	C34	C33	C32	C31	Components of B
C413	C412	C411	C410	C49	C48	C47	C46	C45	C44	C43	C42	C41	Components of B
C513	C512	C511	C510	C59	C58	C57	C56	C55	C54	C53	C52	C51	Components of B
C613	C612	C611	C610	C69	C68	C67	C66	C65	C64	C63	C62	C61	Components of B
C713	C712	C711	C710	C79	C78	C77	C76	C75	C74	C73	C72	C71	Components of B
C813	C812	C811	C810	C89	C88	C87	C86	C85	C84	C83	C82	C81	Components of B
C913	C912	C911	C910	C99	C98	C97	C96	C95	C94	C93	C92	C91	Components of B
C1013	C1012	C1011	C1010	C109	C108	C107	C106	C105	C104	C103	C102	C101	Components of B
C1113	C1112	C1111	C1110	C119	C118	C117	C116	C115	C114	C113	C112	C111	Components of B
C1213	C1212	C1211	C1210	C129	C128	C127	C126	C125	C124	C123	C122	C121	Components of B
C	C	C	C	C	C	C	C	C	C	C	C	C	Component

	1313	1312	1311	1310	139	138	137	136	135	134	133	132	131	s of B
⋮	C_{1413}	C_{1412}	C_{1411}	C_{1410}	C_{149}	C_{148}	C_{147}	C_{146}	C_{145}	C_{144}	C_{143}	C_{142}	C_{141}	Component s of B
⋮	C_{1513}	C_{1512}	C_{1511}	C_{1510}	C_{159}	C_{158}	C_{157}	C_{156}	C_{155}	C_{154}	C_{13}	C_{152}	C_{151}	Component s of B
⋮	C_{1613}	C_{1612}	C_{1611}	C_{1610}	C_{169}	C_{168}	C_{167}	C_{166}	C_{165}	C_{164}	C_{163}	C_{162}	C_{161}	Component s of B
⋮	C_{1713}	C_{1712}	C_{1711}	C_{1710}	C_{179}	C_{178}	C_{177}	C_{176}	C_{175}	C_{174}	C_{173}	C_{172}	C_{171}	Component s of B
⋮	C_{1813}	C_{1812}	C_{1811}	C_{1810}	C_{189}	C_{188}	C_{187}	C_{186}	C_{185}	C_{184}	C_{183}	C_{182}	C_{181}	Component s of B
⋮	C_{1913}	C_{1912}	C_{1911}	C_{1910}	C_{199}	C_{198}	C_{197}	C_{196}	C_{195}	C_{194}	C_{193}	C_{192}	C_{191}	Component s of B
⋮	C_{2013}	C_{2012}	C_{2011}	C_{2010}	C_{209}	C_{208}	C_{207}	C_{206}	C_{205}	C_{204}	C_{203}	C_{202}	C_{201}	Component s of B
⋮	C_{2113}	C_{2112}	C_{2111}	C_{2110}	C_{219}	C_{218}	C_{217}	C_{216}	C_{215}	C_{214}	C_{213}	C_{212}	C_{211}	Component s of B
⋮	C_{2213}	C_{2212}	C_{2211}	C_{2210}	C_{229}	C_{228}	C_{227}	C_{226}	C_{225}	C_{224}	C_{223}	C_{222}	C_{221}	Component s of B
⋮	C_{2313}	C_{2312}	C_{2311}	C_{2310}	C_{239}	C_{238}	C_{237}	C_{236}	C_{235}	C_{234}	C_{233}	C_{232}	C_{231}	Component s of B
⋮	C_{2413}	C_{2412}	C_{2411}	C_{2410}	C_{249}	C_{248}	C_{247}	C_{246}	C_{245}	C_{244}	C_{243}	C_{242}	C_{241}	Component s of B
⋮	C_{2513}	C_{2512}	C_{2511}	C_{2510}	C_{259}	C_{258}	C_{257}	C_{256}	C_{255}	C_{254}	C_{253}	C_{252}	C_{251}	Component s of B
⋮	C_{2613}	C_{2612}	C_{2611}	C_{2610}	C_{269}	C_{268}	C_{267}	C_{266}	C_{265}	C_{264}	C_{263}	C_{262}	C_{261}	Component s of B
⋮	C_{2713}	C_{2712}	C_{2711}	C_{2710}	C_{279}	C_{278}	C_{277}	C_{276}	C_{275}	C_{274}	C_{273}	C_{272}	C_{271}	Component s of B
⋮	C_{2813}	C_{2812}	C_{2811}	C_{2810}	C_{289}	C_{288}	C_{287}	C_{286}	C_{285}	C_{284}	C_{283}	C_{282}	C_{281}	Component s of B
⋮	C_{2913}	C_{2912}	C_{2911}	C_{2910}	C_{299}	C_{298}	C_{297}	C_{296}	C_{295}	C_{294}	C_{293}	C_{292}	C_{291}	Component s of B
⋮	C_{3013}	C_{3012}	C_{3011}	C_{3010}	C_{309}	C_{308}	C_{307}	C_{306}	C_{305}	C_{304}	C_{303}	C_{302}	C_{301}	Component s of B
⋮	C_{3113}	C_{3112}	C_{3111}	C_{3110}	C_{319}	C_{318}	C_{317}	C_{316}	C_{315}	C_{314}	C_{313}	C_{312}	C_{311}	Component s of B
⋮	C_{3213}	C_{3212}	C_{3211}	C_{3210}	C_{329}	C_{328}	C_{327}	C_{326}	C_{325}	C_{324}	C_{323}	C_{322}	C_{321}	Component s of B
⋮	C_{3313}	C_{3312}	C_{3311}	C_{3310}	C_{339}	C_{338}	C_{337}	C_{336}	C_{335}	C_{334}	C_{333}	C_{332}	C_{331}	Component s of B
⋮	C_{3413}	C_{3412}	C_{3411}	C_{3410}	C_{349}	C_{348}	C_{347}	C_{346}	C_{345}	C_{344}	C_{343}	C_{342}	C_{341}	Component s of B
⋮	C_{3513}	C_{3512}	C_{3511}	C_{3510}	C_{359}	C_{358}	C_{357}	C_{356}	C_{355}	C_{354}	C_{353}	C_{352}	C_{351}	Component s of B
⋮	C_{3613}	C_{3612}	C_{3611}	C_{3610}	C_{369}	C_{368}	C_{367}	C_{366}	C_{365}	C_{364}	C_{363}	C_{362}	C_{361}	Component s of B
⋮	C	C	C	C	C	C	C	C	C	C	C	C	C	Component

3713	3712	3711	3710	379	378	377	376	375	374	373	372	371	s of B
C_{3813}	C_{3812}	C_{3811}	C_{3810}	C_{389}	C_{388}	C_{387}	C_{386}	C_{385}	C_{384}	C_{383}	C_{382}	C_{381}	Components of B
C_{3913}	C_{3912}	C_{3911}	C_{3910}	C_{399}	C_{398}	C_{397}	C_{396}	C_{395}	C_{394}	C_{393}	C_{392}	C_{391}	Components of B
C_{4013}	C_{4012}	C_{4011}	C_{4010}	C_{409}	C_{408}	C_{407}	C_{406}	C_{405}	C_{404}	C_{403}	C_{402}	C_{401}	Components of B
C_{4113}	C_{4112}	C_{4111}	C_{4110}	C_{419}	C_{418}	C_{417}	C_{416}	C_{415}	C_{414}	C_{413}	C_{412}	C_{411}	Components of B
C_{4213}	C_{4212}	C_{4211}	C_{4210}	C_{429}	C_{428}	C_{427}	C_{426}	C_{425}	C_{424}	C_{423}	C_{422}	C_{421}	Components of B
C_{4313}	C_{4312}	C_{4311}	C_{4310}	C_{439}	C_{438}	C_{437}	C_{436}	C_{435}	C_{434}	C_{433}	C_{432}	C_{431}	Components of B
C_{4413}	C_{4412}	C_{4411}	C_{4410}	C_{449}	C_{448}	C_{447}	C_{446}	C_{445}	C_{444}	C_{443}	C_{442}	C_{441}	Components of B
C_{4513}	C_{4512}	C_{4511}	C_{4510}	C_{459}	C_{458}	C_{457}	C_{456}	C_{455}	C_{454}	C_{453}	C_{452}	C_{451}	Components of B
C_{4613}	C_{4612}	C_{4611}	C_{4610}	C_{469}	C_{468}	C_{467}	C_{466}	C_{465}	C_{464}	C_{463}	C_{462}	C_{461}	Components of B
C_{4713}	C_{4712}	C_{4711}	C_{4710}	C_{479}	C_{478}	C_{477}	C_{476}	C_{475}	C_{474}	C_{473}	C_{472}	C_{471}	Components of B
C_{4813}	C_{4812}	C_{4811}	C_{4810}	C_{489}	C_{488}	C_{487}	C_{486}	C_{485}	C_{484}	C_{483}	C_{482}	C_{481}	Components of B
C_{4913}	C_{4912}	C_{4911}	C_{4910}	C_{499}	C_{498}	C_{497}	C_{496}	C_{495}	C_{494}	C_{493}	C_{492}	C_{491}	Components of B
C_{5013}	C_{5012}	C_{5011}	C_{5010}	C_{509}	C_{508}	C_{507}	C_{506}	C_{505}	C_{504}	C_{503}	C_{502}	C_{501}	Components of B
C_{5113}	C_{5112}	C_{5111}	C_{5110}	C_{519}	C_{518}	C_{517}	C_{516}	C_{515}	C_{514}	C_{513}	C_{512}	C_{511}	Components of B
C_{5213}	C_{5212}	C_{5211}	C_{5210}	C_{529}	C_{528}	C_{527}	C_{526}	C_{525}	C_{524}	C_{523}	C_{522}	C_{521}	Components of B
C_{5313}	C_{5312}	C_{5311}	C_{5310}	C_{539}	C_{538}	C_{537}	C_{536}	C_{535}	C_{534}	C_{533}	C_{532}	C_{531}	Components of B
C_{5413}	C_{5412}	C_{5411}	C_{5410}	C_{549}	C_{548}	C_{547}	C_{546}	C_{545}	C_{544}	C_{543}	C_{542}	C_{541}	Components of B
C_{5513}	C_{5512}	C_{5511}	C_{5510}	C_{559}	C_{558}	C_{557}	C_{556}	C_{555}	C_{554}	C_{553}	C_{552}	C_{551}	Components of B
C_{5613}	C_{5612}	C_{5611}	C_{5610}	C_{569}	C_{568}	C_{567}	C_{566}	C_{565}	C_{564}	C_{563}	C_{562}	C_{561}	Components of B
C_{5713}	C_{5712}	C_{5711}	C_{5710}	C_{579}	C_{578}	C_{577}	C_{576}	C_{575}	C_{574}	C_{573}	C_{572}	C_{571}	Components of B
C_{5813}	C_{5812}	C_{5811}	C_{5810}	C_{589}	C_{588}	C_{587}	C_{586}	C_{585}	C_{584}	C_{583}	C_{582}	C_{581}	Components of B
C_{5913}	C_{5912}	C_{5911}	C_{5910}	C_{599}	C_{598}	C_{597}	C_{596}	C_{595}	C_{594}	C_{593}	C_{592}	C_{591}	Components of B
C_{6013}	C_{6012}	C_{6011}	C_{6010}	C_{609}	C_{608}	C_{607}	C_{606}	C_{605}	C_{604}	C_{603}	C_{602}	C_{601}	Components of B

:													
C 6113	C 6112	C 6111	C 6110	C 619	C 618	C 617	C 616	C 615	C 614	C 613	C 612	C 611	Components of B
C 6213	C 6212	C 6211	C 6210	C 629	C 628	C 627	C 626	C 625	C 624	C 623	C 622	C 621	Components of B
C 6313	C 6312	C 6311	C 6310	C 639	C 638	C 637	C 636	C 635	C 634	C 633	C 632	C 631	Components of B
C 6413	C 6412	C 6411	C 6410	C 649	C 648	C 647	C 646	C 645	C 644	C 643	C 642	C 641	Components of B
C 6513	C 6512	C 6511	C 6510	C 659	C 658	C 657	C 656	C 655	C 654	C 653	C 652	C 651	Components of B
C 6613	C 6612	C 6611	C 6610	C 669	C 668	C 667	C 666	C 665	C 664	C 663	C 662	C 661	Components of B
C 6713	C 6712	C 6711	C 6710	C 679	C 678	C 677	C 676	C 675	C 674	C 673	C 672	C 671	Components of B
C 6813	C 6812	C 6811	C 6810	C 689	C 688	C 687	C 686	C 685	C 684	C 683	C 682	C 681	Components of B
C 6913	C 6912	C 6911	C 6910	C 699	C 698	C 697	C 696	C 695	C 694	C 693	C 692	C 691	Components of B
C 7013	C 7012	C 7011	C 7010	C 709	C 708	C 707	C 706	C 705	C 704	C 703	C 702	C 701	Components of B
C 7113	C 7112	C 7111	C 7110	C 719	C 718	C 717	C 716	C 715	C 714	C 713	C 712	C 711	Components of B
C 7213	C 7212	C 7211	C 7210	C 729	C 728	C 727	C 726	C 725	C 724	C 723	C 722	C 721	Components of B
C 7313	C 7312	C 7311	C 7310	C 739	C 738	C 737	C 736	C 735	C 734	C 733	C 732	C 731	Components of B
C 7413	C 7412	C 7411	C 7410	C 749	C 748	C 747	C 746	C 745	C 744	C 743	C 742	C 741	Components of B
C 7513	C 7512	C 7511	C 7510	C 759	C 758	C 757	C 756	C 755	C 754	C 753	C 752	C 751	Components of B
C 7613	C 7612	C 7611	C 7610	C 769	C 768	C 767	C 766	C 765	C 764	C 763	C 762	C 761	Components of B
C 7713	C 7712	C 7711	C 7710	C 779	C 778	C 777	C 776	C 775	C 774	C 773	C 772	C 771	Components of B
C 7813	C 7812	C 7811	C 7810	C 789	C 788	C 787	C 786	C 785	C 784	C 783	C 782	C 781	Components of B
C 7913	C 7912	C 7911	C 7910	C 799	C 798	C 797	C 796	C 795	C 794	C 793	C 792	C 791	Components of B
C 8013	C 8012	C 8011	C 8010	C 809	C 808	C 807	C 806	C 805	C 804	C 803	C 802	C 801	Components of B
C 8113	C 8112	C 8111	C 8110	C 819	C 818	C 817	C 816	C 815	C 814	C 813	C 812	C 811	Components of B
C 8213	C 8212	C 8211	C 8210	C 829	C 828	C 827	C 826	C 825	C 824	C 823	C 822	C 821	Components of B
C 8313	C 8312	C 8311	C 8310	C 839	C 838	C 837	C 836	C 835	C 834	C 833	C 832	C 831	Components of B
C 8413	C 8412	C 8411	C 8410	C 849	C 848	C 847	C 846	C 845	C 844	C 843	C 842	C 841	Components of B

⋮	C 8513	C 8512	C 8511	C 8510	C 859	C 858	C 857	C 856	C 855	C 854	C 853	C 852	C 851	Components of B
⋮	C 8613	C 8612	C 8611	C 8610	C 869	C 868	C 867	C 866	C 865	C 864	C 863	C 862	C 861	Components of B
⋮	C 8713	C 8712	C 8711	C 8710	C 879	C 878	C 877	C 876	C 875	C 874	C 873	C 872	C 871	Components of B
⋮	C 8813	C 8812	C 8811	C 8810	C 889	C 888	C 887	C 886	C 885	C 884	C 883	C 882	C 881	Components of B
⋮	C 8913	C 8912	C 8911	C 8910	C 899	C 898	C 897	C 896	C 895	C 894	C 893	C 892	C 891	Components of B
⋮	C 9013	C 9012	C 9011	C 9010	C 909	C 908	C 907	C 906	C 905	C 904	C 903	C 902	C 901	Components of B
⋮	C 9113	C 9112	C 9111	C 9110	C 919	C 918	C 917	C 916	C 915	C 914	C 913	C 912	C 911	Components of B
⋮	C 9213	C 9212	C 9211	C 9210	C 929	C 928	C 927	C 926	C 925	C 924	C 923	C 922	C 921	Components of B
⋮	C 9313	C 9312	C 9311	C 9310	C 939	C 938	C 937	C 936	C 935	C 934	C 933	C 932	C 931	Components of B
⋮	C 9413	C 9412	C 9411	C 9410	C 949	C 948	C 947	C 946	C 945	C 944	C 943	C 942	C 941	Components of B
⋮	C 9513	C 9512	C 9511	C 9510	C 959	C 958	C 957	C 956	C 955	C 954	C 953	C 952	C 951	Components of B
⋮	C 9613	C 9612	C 9611	C 9610	C 969	C 968	C 967	C 966	C 965	C 964	C 963	C 962	C 961	Components of B
⋮	C 9713	C 9712	C 9711	C 9710	C 979	C 978	C 977	C 976	C 975	C 974	C 973	C 972	C 971	Components of B
⋮	C 9813	C 9812	C 9811	C 9810	C 989	C 988	C 987	C 986	C 985	C 984	C 983	C 982	C 981	Components of B
⋮	C 9913	C 9912	C 9911	C 9910	C 999	C 998	C 997	C 996	C 995	C 994	C 993	C 992	C 991	Components of B
⋮	C 10013	C 10012	C 10011	C 10010	C 1009	C 1008	C 1007	C 1006	C 1005	C 1004	C 1003	C 1002	C 1001	Components of B
⋮	C 10113	C 10112	C 10111	C 10110	C 1019	C 1018	C 1017	C 1016	C 1015	C 1014	C 1013	C 1012	C 1011	Components of B
⋮	C 10213	C 10212	C 10211	C 10210	C 1029	C 1028	C 1027	C 1026	C 1025	C 1024	C 1023	C 1022	C 1021	Components of B
⋮	C 10313	C 10312	C 10311	C 10310	C 1039	C 1038	C 1037	C 1036	C 1035	C 1034	C 1033	C 1032	C 1031	Components of B
⋮	C 10413	C 10412	C 10411	C 10410	C 1049	C 1048	C 1047	C 1046	C 1045	C 1044	C 1043	C 1042	C 1041	Components of B
⋮	C 10513	C 10512	C 10511	C 10510	C 1059	C 1058	C 1057	C 1056	C 1055	C 1054	C 1053	C 1052	C 1051	Components of B
⋮	C 10613	C 10612	C 10611	C 10610	C 1069	C 1068	C 1067	C 1066	C 1065	C 1064	C 1063	C 1062	C 1061	Components of B
⋮	C 10713	C 10712	C 10711	C 10710	C 1079	C 1078	C 1077	C 1076	C 1075	C 1074	C 1073	C 1072	C 1071	Components of B
⋮	C	C	C	C	C	C	C	C	C	C	C	C	C	Component

	10813	10812	10811	10810	1089	1088	1087	1086	1085	1084	1083	1082	1081	s of B
⋮	C_{10913}	C_{10912}	C_{10911}	C_{10910}	C_{1099}	C_{1098}	C_{1097}	C_{1096}	C_{1095}	C_{1094}	C_{1093}	C_{1092}	C_{1091}	Components of B
⋮	C_{11013}	C_{11012}	C_{11011}	C_{11010}	C_{1109}	C_{1108}	C_{1107}	C_{1106}	C_{1105}	C_{1104}	C_{1103}	C_{1102}	C_{1101}	Components of B
⋮	C_{11113}	C_{11112}	C_{11111}	C_{11110}	C_{1119}	C_{1118}	C_{1117}	C_{1116}	C_{1115}	C_{1114}	C_{1113}	C_{1112}	C_{1111}	Components of B
⋮	C_{11213}	C_{11212}	C_{11211}	C_{11210}	C_{1129}	C_{1128}	C_{1127}	C_{1126}	C_{1125}	C_{1124}	C_{1123}	C_{1122}	C_{1121}	Components of B
⋮	C_{11313}	C_{11312}	C_{11311}	C_{11310}	C_{1139}	C_{1138}	C_{1137}	C_{1136}	C_{1135}	C_{1134}	C_{1133}	C_{1132}	C_{1131}	Components of B
⋮	C_{11413}	C_{11412}	C_{11411}	C_{11410}	C_{1149}	C_{1148}	C_{1147}	C_{1146}	C_{1145}	C_{1144}	C_{1143}	C_{1142}	C_{1141}	Components of B
⋮	C_{11513}	C_{11512}	C_{11511}	C_{11510}	C_{1159}	C_{1158}	C_{1157}	C_{1156}	C_{1155}	C_{1154}	C_{1153}	C_{1152}	C_{1151}	Components of B
⋮	C_{11613}	C_{11612}	C_{11611}	C_{11610}	C_{1169}	C_{1168}	C_{1167}	C_{1166}	C_{1165}	C_{1164}	C_{1163}	C_{1162}	C_{1161}	Components of B
⋮	C_{11713}	C_{11712}	C_{11711}	C_{11710}	C_{1179}	C_{1178}	C_{1177}	C_{1176}	C_{1175}	C_{1174}	C_{1173}	C_{1172}	C_{1171}	Components of B
⋮	C_{11813}	C_{11812}	C_{11811}	C_{11810}	C_{1189}	C_{1188}	C_{1187}	C_{1186}	C_{1185}	C_{1184}	C_{1183}	C_{1182}	C_{1181}	Components of B
⋮	C_{11913}	C_{11912}	C_{11911}	C_{11910}	C_{1199}	C_{1198}	C_{1197}	C_{1196}	C_{1195}	C_{1194}	C_{1193}	C_{1192}	C_{1191}	Components of B
⋮	C_{12013}	C_{12012}	C_{12011}	C_{12010}	C_{1209}	C_{1208}	C_{1207}	C_{1206}	C_{1205}	C_{1204}	C_{1203}	C_{1202}	C_{1201}	Components of B
⋮	C_{12113}	C_{12112}	C_{12111}	C_{12110}	C_{1219}	C_{1218}	C_{1217}	C_{1216}	C_{1215}	C_{1214}	C_{1213}	C_{1212}	C_{1211}	Components of B
⋮	C_{12213}	C_{12212}	C_{12211}	C_{12210}	C_{1229}	C_{1228}	C_{1227}	C_{1226}	C_{1225}	C_{1224}	C_{1223}	C_{1222}	C_{1221}	Components of B
⋮	C_{12313}	C_{12312}	C_{12311}	C_{12310}	C_{1239}	C_{1238}	C_{1237}	C_{1236}	C_{1235}	C_{1234}	C_{1233}	C_{1232}	C_{1231}	Components of B
⋮	C_{12413}	C_{12412}	C_{12411}	C_{12410}	C_{1249}	C_{1248}	C_{1247}	C_{1246}	C_{1245}	C_{1244}	C_{1243}	C_{1242}	C_{1241}	Components of B
⋮	C_{12513}	C_{12512}	C_{12511}	C_{12510}	C_{1259}	C_{1258}	C_{1257}	C_{1256}	C_{1255}	C_{1254}	C_{1253}	C_{1252}	C_{1251}	Components of B
⋮	C_{12613}	C_{12612}	C_{12611}	C_{12610}	C_{1269}	C_{1268}	C_{1267}	C_{1266}	C_{1265}	C_{1264}	C_{1263}	C_{1262}	C_{1261}	Components of B
⋮	C_{12713}	C_{12712}	C_{12711}	C_{12710}	C_{1279}	C_{1278}	C_{1277}	C_{1276}	C_{1275}	C_{1274}	C_{1273}	C_{1272}	C_{1271}	Components of B
⋮	C_{12813}	C_{12812}	C_{12811}	C_{12810}	C_{1289}	C_{1288}	C_{1287}	C_{1286}	C_{1285}	C_{1284}	C_{1283}	C_{1282}	C_{1281}	Components of B
⋮	C_{12913}	C_{12912}	C_{12911}	C_{12910}	C_{1299}	C_{1298}	C_{1297}	C_{1296}	C_{1295}	C_{1294}	C_{1293}	C_{1292}	C_{1291}	Components of B
⋮	.	.	.	.	.	.	.	.	.	.	.	.	.	.
	.	.	.	.	.	.	.	.	.	.	.	.	.	.

anything you see in table 3-1 is same equations that you saw in Chapters one and two. The author in the early part of his research evolution called these types of studies as "standard studies". And appropriate methodology with it carved "standardized methodology of the research" or even "evolutionary methodology". The research methodology is fully presented in the 2014 paper published in the University and record at the National Elite Foundation of Iran. The author of the present study, due to the inventor this method in 2014, names these studies in the situtaion words form of the "Research (Cuonter) Component", in numerical form "number Component" and in the general case "Creative thinking on visible surfaces" and It naming and carved.

References

C. Montgomery D. (2012); Design and Analysis of Experiments, Arizona State University, Eighth Edition, PP. 1-757.

C. Montgomery D. (2009); Introduction to Statistical Quality Control, Arizona State University, Six Edition, PP. 62-112.

C. Montgomery D. (2002); Applied Statistics and Probability for Engineers, Arizona State University, Third Edition, PP. 221.

C. Montgomery D. (2001); Design and Analysis of Experiments (Translation by Gholamhossein Shahkar), First Edition, University Publication Center, University of Tehran, PP.1-500.

Shahkar Gh.H. (1996); Design Experiments, First Edition, Publication Payam Noor University, Payam Noor University.

Shahkar Gh.H. (1999); Design Experiments 2, First Edition, Publication Payam Noor University, Payam Noor University, PP.1-131.

Valiollahpur M. (2019); Management Components for Creativity and Innovation, LAP Lambert Academic Publishing. At Amazon: https://www.amazon.it/Management-Components-Creativity-Innovation-Scientific/dp/3659648329.

Valiollahpur M. (2017); The Analysis of International Trade in the Development of Economic Ties based on the Geopolitical Approach, Ariya-Dansh Publishing. At: http://opac.nlai.ir/opac-prod/bibliographic/5112722.

Valiollahpur M. (2017); Presenting a Knowledge-based Innovation Model, Ariya-Dansh Publishing. At: http://opac.nlai.ir/opac-prod/bibliographic/5112722.

Valiollahpur M. (2017); The Effects of Creativity and Innovation Components on Organizational Performance Development, Bartar-Andishan Publishing. At: http://opac.nlai.ir/opac-prod/bibliographic/5289329.

Valiollahpur, M. & Valiollahpur, S. (2014); *"Studying Commercial Metropol on International Economic Policies in The Islamic Republic of Iran's Mutual Interaction Flows by Country Separation with Analytic Attitude of Geoinformatics and Geomatics"*, National conference on economics and management, favorite presentation, published in the journal of association of social science research, July 2014. At Available (Academic Elite Paper): https://sina.bmn.ir.

Post-Portrait

For all of the issues that you are designing, the revealed structures will justify most organized and unstructured methods that it will show you the right policies for using information. How when that you wish assess the target a micro level for example for an organization or at a macro level for example for the community to give the best result to yourself, but when you use your essential information, only the result is enough, in other words, do not include them in your attitude systems the data that references the necessary information to you does not bring you trusts. it may also show deviations from your values so that you think moving with you. This subject goes beyond the scientific process of countries and when it becomes clear that you want with all sorts of your reason to engaged in analyze the complexities of science in ways beyond of linear and nonlinear thinking.

برای تمام موضوعاتی که شما در حال طراحی هستید، ساختارهای نشان داده شده توجیه اکثر روش‌های سازمان‌یافته و سازمان‌نیافته ا ت که مسیر یا تهای صحیح ا تفاده از اطلاعات را آشکار خواهد ساخت. چگونه شما هنگامی که می‌خواهید هدفی را در طح خرد مثلاً برای یک سازمان یا در طح کلان مثلاً برای جامعه ارزیابی نمایید از جزئی‌ترین موارد نمی‌گذرید تا نتیجه را به بهترین شکل برای خود رقم بزنید ولی هنگامی که از اطلاعات ضروری خود ا تفاده می‌کنید فقط به نتیجه آن بسنده می‌کنید. برای مسائلی که شما طرح می‌کنید نیز ساختارها همین‌گونه هستند تا این موارد را در سیستم‌های نگرشی خود لحاظ ننمایید داده‌هایی که اطلاعات ضروری را به شما ارجاع می‌دهند اعتمادی را برای شما به ارمغان نمی‌آورند. شاید هم انحراف از ارزش‌های شما را به صورتی برایتان نشان دهند که شما فکر می‌کنید هم‌را تا با شما در حرکتند. این موضوع فراتر از روند علمی کشورها ا ت و زمانی اهمیت خود را آشکار می‌کند که شما بخواهید با انواع و اقسام دلایل خود پیچیدگی‌های علم را در بسطهای بالاتر از تفکر خطی و غیر خطی تحلیل کنید.

In order to explain plans, strategies, experimental research methods, strategies, uncertainties ... (from Theories, Philosophies, Explanations, Effects, Aftereffects, Comparisons, Frameworks, Models, Share, Reasons, The share of aid, ... to Phenomenology, Ontology, Epistemology, and ... as a small part of research) the important new thinking in the research method of sciences. The scientific breakthrough is a situation formed on the perception of perimeter of a researcher who has a insight. This status show self face in trade, invest, culture, politics, economics, society and etc. A method of science that to it has been given a appropriate answer in the past 50 decades in international scientific societies, but has not been given a response to its evolution. It's mean in that we should thinking and understand the true science of the research. Looking at the developmental indicators the entry into the new era placed structures in self that have embodied non-realistic practices with the principles of realistic, yet at with simple outward but complex inward that can not be easily overlooked and these are not just the principles of mathematics or the basic fundamentals of knowledge.

ها، روش پژوهش‌های به منظور تبیین برنامه‌ریزی‌ها، ا ستراتژی آزمایشی، راهبردها، عدم قطعیت‌ها و ... (از نظریه‌ها، فلسفه‌ها، تبیین‌ها، اثرات، تبعات، مقایسه‌ها، چارچوب‌ها، مدل‌ها، هم‌بندی‌ها، دلیل‌ها، همیاری‌ها و ... گرفته تا پدیدارشنا ی، هستی‌شنا ی، معرفتشنا ی و ... به عنوان گوشه‌ای از تحقیقات) تفکر جدید در روش تحقیق علوم مهم می‌باشد. شکاف (مرز شکنی) علمی وضعیتی ا ت که بر روی درک محیط پیرامون محققی که دارای بصیرت ا ت شکل می گیرد. این وضعیت خود را در تجارت، رمایه گذاری، فرهنگ، یا ت، اقتصاد، جامعه و غیره نشان می‌دهد. یک روش علمی که در صده‌ها و دهه‌های گذشته در جوامع علمی بین المللی جواب تنسیق شده‌ای در جوامع علمی بین‌الملل به آن داده شده ا ت اما جوابی برای تکامل آن داده نشده ا ت. این بدان معنی ا ت که ما باید فکر کنیم و علم واقعی پژوهش را درک نماییم. با نگاهی به شاخص‌های پیشرفت (تو عه‌یافتگی)، وارد شدن به دوره جدید ساختارهایی را در خود جای داده ا ت که رویه‌های غیر واقعی را با اصول پایه‌ریزی‌شده و تجزیه‌پذیر و در عین حال با ظاهری ملهوف اما باطنی ملفوف همراه کرده ا ت که نمی‌توان به سادگی از کنار آن گذشت و صرفاً این اصول، ریاضیات یا مبانی دانش پایه نیستند.

Important: This statement for the first time was presented by Dr. Morteza Valiollahpur in April 2018 at the Iranian National Foundation of Elites at: http://en.bmn.ir.